Help Teens Overcome *Math Anxiety*

For Teachers, Students, & Parents:

5 Strategies to My Success

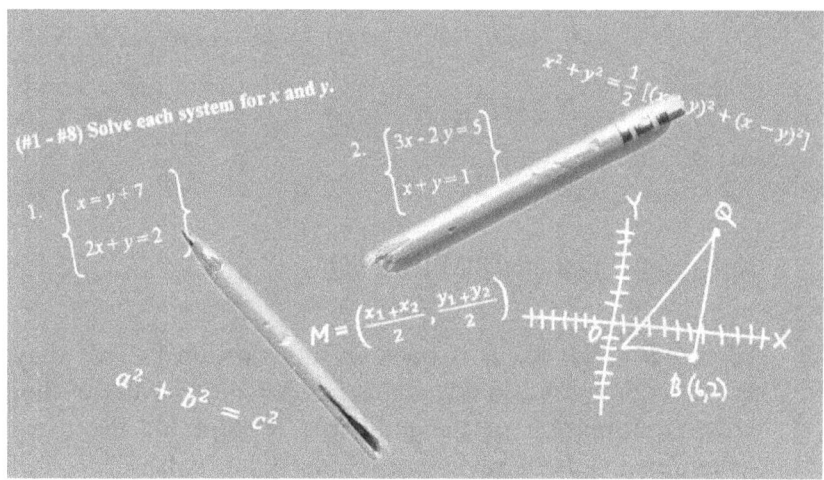

Dianne DeMille, PhD

It's Not About How You Teach, It's About How Your Students Learn

Copyright © Dianne DeMille, PhD

All rights reserved. This book or any portion thereof may not be reproduced or used in any manner whatsoever without the express written permission of the publisher, except for the use of brief quotations in a book review or scholarly journal.

October 2025
Dianne's Consultant Services
Fullerton, CA

Paperback ISBN: 979-8-9992321-8-2

Help Teens Overcome Math Anxiety

Practice – Self-Defeating Attitudes = Success • Success • Success ...

Help Teens Overcome Math Anxiety

For Teachers, Students, & Parents:
5 Strategies to My Success

Introduction

Teens can be very emotional and may feel intimidated when they don't understand something. They often avoid asking questions or seek support from their teacher or peers. Many have self-doubt, fear of failure and judgment by others, creating a cycle that increases anxiety about doing math. This book offers strategies to help change their mindset.

As a math educator for over 50 years, I've witnessed countless students struggle with the fear and frustration of learning math. I vividly remember one student, Sarah, who would tremble at the mere thought of solving equations. It broke my heart to see her potential being held back by this overwhelming emotion.

I want to share my rewarding experience teaching high school mathematics over the past few decades. Students appreciated discussions about math anxiety and journaling in my classes. Specific notes for parents are included in the last chapter.

In the summer of 1982, after a decade of teaching math, I attended a six-week professional development opportunity for secondary math teachers at UCLA. One assignment changed everything for me. We were asked to explore an area of mathematics and share what we learned.

I stumbled across *Overcoming Math Anxiety* by Sheila Tobias. Her book instantly grabbed my attention. It was about fear; real, paralyzing fear that kept good students, especially women, from ever discovering their mathematical potential.[i]

Tobias's work stemmed from an article she wrote for *Ms. Magazine* in 1976. Gloria Steinem was a driving force in the women's movement and co-founder of *Ms. Magazine*. She called the article "one of the most important pieces we've ever published.[ii] Back then, societal expectations discouraged girls from pursuing higher-level math. Tobias exposed a hidden epidemic: nine out of ten women suffered from math anxiety. And she was determined to diminish the epidemic.[iii]

The River of Knowledge

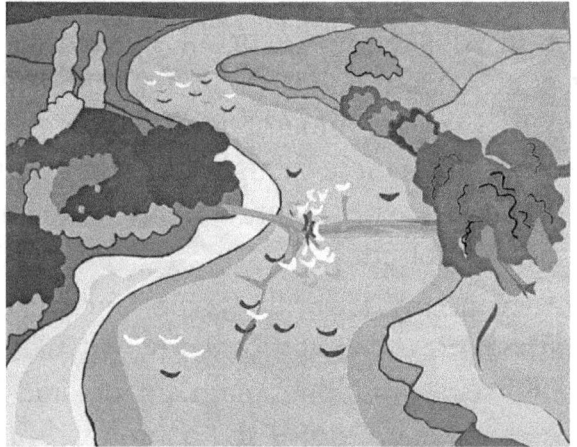

I also read about "The River of Knowledge" and how we learn. The concept is presented to us as a metaphor for the continuous flow of wisdom from various cultural and religious traditions. The river flows smoothly as we begin our learning journey. We pick up pebbles of information along the way. This process encompasses all learning as

Help Teens Overcome Math Anxiety

Practice – Self-Defeating Attitudes = Success • Success • Success ...

we navigate various obstacles. We may approach larger pebbles or boulders that prompt us to stop and think more deeply about how we might meander around them to continue our journey. Sometimes, we may encounter a large boulder or a fallen tree that blocks our path. This roadblock in our learning requires special attention and thought to help maneuver around it. Once we overcome that obstacle, the river opens, allowing us to continue learning.

Most rivers have detours or roadblocks and for some students, their roadblocks are early in the learning process, while others may face them later. These roadblocks to learning math, or learning in general, occur at different times for each person. Whether it's division, fractions, algebra, or at some other point when things become more complex. Students can learn how to break through their roadblocks. Once they address their difficulties and discover ways to navigate those problems, they unlock their "river of knowledge" toward success.

When I looked back at my own mathematical journey, I realized I had stumbled a couple of times.

Example: My Fist Roadblock

> In fourth grade, the teacher placed me in a slow math group. I was surprised and hurt because I had always believed I was a good student. She commented that I sometimes went too quickly and made minor errors.
>
> I decided to study hard and prove to her that I deserved to be moved up. Eventually, she

> recognized my progress and moved me to a different group. This was my first roadblock.

Example: Calculus

> Another example was when I took Calculus in college. I was determined to find a way around the roadblock. It took me some time to realize I needed to study more and might not get the answers as quickly as I did in high school. I needed to invest time understanding the depth of the problems and how they functioned to recognize I could succeed in doing the math. Maintaining this positive mindset became most effective for me.

I want all students to visualize how they can support themselves by eliminating self-defeating attitudes that hold them back. All students could see the following equation in large font above my chalkboard: **"Practice − Self-Defeating Attitudes = Success × Success × Success ..."** and they would often point it out or repeat it during class.

Why I Wrote This Book

Tobias's message impressed me as I found many of my students, both male and female, struggled to learn math. I shared my experience at many math conferences throughout California and nationally to help teachers assist their students to overcome math anxiety using Tobia's advice, the river of knowledge, and the success I achieved with my students.

Beginning in the fall of 1982, I shared with my students each year in every math course the strategies I learned and

Help Teens Overcome Math Anxiety

Practice – Self-Defeating Attitudes = Success • Success • Success ...

integrated the "river of knowledge." These strategies are included in this book.

Example: Discuss Math Anxiety

> During the first week of school, I discussed math anxiety and some ways to overcome it. I wanted my students to have some tools to use throughout the course.
>
> I shared how I experienced anxiety when I started calculus in college and how understanding the "River of Knowledge" supported my learning.

Throughout the year, I referred to our initial discussion about math anxiety and encouraged students to write in their journals about their own experiences. Many journal questions are included at various points throughout this book. You may have other questions of your own. My hope is that you will use them regularly with your students throughout your classes.

Sarah's story is not unique. In 2019, 31% of adolescents claimed, "I can't do math!" This widespread math anxiety is a significant barrier to academic success and personal growth, and we must address it head-on. Why is it socially acceptable to be poor at doing math? The International Piza Report compared changes in math anxiety for several countries. The United States went from 29% in 2012 to 38% in 2022 of students who experience math anxiety. A similar trend was found in most countries.[iv]

Through my decades of teaching secondary mathematics and currently mentoring student teachers, I've developed a keen understanding of what works. I've seen firsthand how the right strategies can transform student relationships with math, and opening new possibilities for their future.

Help Teens Overcome Math Anxiety

Practice – Self-Defeating Attitudes = Success • Success • Success ...

Multiple students came to me in later years, while taking advanced courses or after beginning college.

For example: John said to me, "You know, when you talked with us about overcoming math anxiety, you helped me see that I could do the math. I began to enjoy it and got better grades. Thank you for helping me!"

Marianna said, "I can't believe how well I'm doing in my Algebra II class! I learned so much from you. Before we talked about getting over our math anxiety, I didn't do well on tests. Now, I'm getting all A's and B's! Thanks for all your help!"

This is why I wrote this book. I want to share the five strategies I found to be significantly effective: mind shifts for success, mathematical practices, problem solving and critical thinking, reflection and reinforcement, and interventions. When implemented consistently, they boosted student engagement, improved test scores, and most importantly, helped students develop a positive mindset toward learning and doing mathematics.

Example: A Beaming Smile

> One year after working to conquer math anxiety in my class, Sarah came to me with a beaming smile.
>
> "I did it," she said, holding up a perfect score on her math test. "I finally understood what I was doing, and it clicked." That moment of triumph, that shift **from anxiety to confidence**, is what I want for every student who struggles.

Throughout this book, I share the practical techniques I have used with my students and the insights I gained over the years from research and my experience. This transformation doesn't happen overnight. It requires a commitment from educators, students, and alike to fully engage with the strategies and create a parents culture of support and encouragement. It means celebrating the small victories along the way and reminding students that their struggles do not define their abilities.

You Will:

- ☐ Receive a set of strategies to assist your students in overcoming math anxiety to unlock their full potential.
- ☐ Discover how to foster a supportive learning environment that motivates students to take risks, embrace challenges, and develop a growth mindset.
- ☐ Observe enhancements in students' mathematics abilities, self-assurance, and overall academic achievements.
- ☐ Encourage students to approach math with curiosity and excitement instead of fear and dread.

By reflecting on our experiences, we can better empathize with our students and cultivate a learning environment that encourages growth, resilience, and a passion for mathematics. So, let's embark on this transformative journey together, one day at a time.

Before we begin, I invite you to consider your own personal journey in mathematics.

Teacher's Journal

Help Teens Overcome Math Anxiety

Practice – Self-Defeating Attitudes = Success • Success • Success ...

> - When have you faced a challenging problem and persevered?
> - Who has influenced your math learning, and how have they shaped your approach?
> - What does math success look like for you?

Let's embark on this journey together!

Table of Contents

Introduction _____ *iii*

Table of Contents _____ *xii*

Chapter 1: Understanding Math Anxiety _____ 14
- Math Myths _____ 14
- The Psychology Behind Math Anxiety _____ 15
- Recognizing the Symptoms in Adolescents _____ 17

Chapter 2: The 5 Strategies _____ 24
- Starting off on the right foot! _____ 24
- 2.1 Mind Shifts for Success _____ 27
- 2.2 Mathematical Practices _____ 35
- 2.3 Problem-Solving and Critical-Thinking ____ 46
- 2.4 Mathematics Interventions _____ 54
- 2.5 Reflection and Reinforcement _____ 59

Chapter 3: Educators & Parents _____ 63
- 3.1 Change Agents _____ 63
- 3.2 Encourage At-Home Math Activities _____ 66

Conclusion _____ 71

Acknowledgements _____ 75

About the Author _____ 77

References _____ 79

Help Teens Overcome Math Anxiety

Practice − Self-Defeating Attitudes = Success • Success • Success ...

Chapter 1:
Understanding Math Anxiety

Have you ever found yourself staring at a math problem, feeling that familiar knot tighten in your stomach? You're not alone. Math anxiety is a genuine phenomenon affecting many adults and nearly one-third of adolescents, causing them to conclude, "I can't do math!"[v] I've observed this painful cycle over the years.

This chapter focuses on understanding the origins of these fears and their manifestations.

Math Myths

When I recently shared with adults about writing this book, their responses were the usual; "I never understood math." "I was no good at math." "Is there really such a thing as 'math anxiety'?" "I never understood it." "I never liked doing math." "Is there hope for me to not be afraid of doing things with numbers?"

I encouraged them to take on a positive mindset and take things slowly. If they can believe they can do it, then they will most likely do just fine.

In *Mind Over Math,* Stanley Kogelman and Joseph Warren identified "The 12 Math Myths."[vi] David Gay stated in *Solving Problems Using Elementary Mathematics,* "Perhaps a better name for these 'myths' would be the '12 Math Misconceptions!'"[vii] And, most recently, Jan 21, 2017, in the article *Math Myths:*

Help Teens Overcome Math Anxiety

Practice – Self-Defeating Attitudes = Success • Success • Success ...

Researchers Debunk Common Misconceptions, researchers at Peabody found "there are many ways math is learned and [teachers] are developing innovative new ways to teach it. They believe that math is not an unyielding discipline, accessible to only a select few. And, they would argue, math is fun."[viii] The following misconceptions were researched: "not everyone is capable of learning math ... math is only about memorization ... math isn't supposed to be fun ... math is something you do alone ... preschoolers are too young to learn math ... math is for boys."[ix]

The Psychology Behind Math Anxiety

The origins of math anxiety are as varied as they are complex. Historically, people did not widely recognize it as a distinct psychological issue. However, modern research has given us further insight. The term "math anxiety" gained traction in the late 20th century, as educators and psychologists observed patterns among students who excelled in school but struggled with math.

The fear of failure, a powerful psychological force, plays a significant role. Many students internalize early experiences of struggle as evidence of their incapacity. They fail to realize that these are merely steppingstones to understanding.

Cognitive distortions, such as black-and-white thinking—seeing themselves as either a math person or not—intensifies this fear. These distortions change perceptions, making challenges seem insurmountable, and reinforce the belief that failure is inevitable.

Research shows that math anxiety can activate the same regions of the brain associated with physical pain, such as the insula and the amygdala, that are responsible for fear responses. Solving a math problem can feel as intimidating as standing on a stage under a spotlight.

Societal influences play a crucial role in the development and perpetuation of math anxiety. For example, gender stereotypes imply that boys are inherently better at math than girls, even though evidence shows otherwise.[x]

Individual predispositions also contribute to math anxiety, making some adolescents more susceptible than others. Perfectionism, for example, is a double-edged sword. While it drives some students to excel, for others, the fear of making mistakes can be paralyzing. These students equate their worth with their ability to achieve perfect scores, leading to immense pressure and anxiety.

Students with low self-efficacy often avoid math, choosing instead to focus on subjects where they feel more confident. By avoiding math, they become more convinced they're bad at it, and this becomes a destructive loop.

Help Teens Overcome Math Anxiety

Practice – Self-Defeating Attitudes = Success • Success • Success ...

Recognizing the Symptoms in Adolescents

By identifying the factors contributing to math anxiety, we can dismantle them, replacing fear with **understanding** and anxiety with **confidence**. As we progress through this book, we will explore strategies and interventions to help students like Tom and many others navigate their math experiences with **greater confidence** and less fear. Together, we can reframe math not as a daunting challenge but as an **exciting opportunity for growth and discovery**.

> As a math teacher, I have witnessed students freeze during tests, unable to recall the information they studied. It's not that they didn't understand the material; their brains were overwhelmed by their emotional response to stress. This physiological reaction highlights the importance of strategies to reduce anxiety and enhance cognitive efficiency.

Imagine a typical classroom setting where students have their eyes fixed on the clock, awaiting the dreaded timed test. The weight of the ticking seconds is intense; each tick serves as a reminder of how little time remains to solve the problems on the page. For many students, this scenario is a pressure cooker for anxiety.

Example: Test

> Tom, for example, a bright student with a curious mind. He thrived in science and language arts but froze at the sight of numbers. One day, he approached me, visibly shaken after a test. He confessed the numbers seemed to blur, his mind went blank, and he couldn't remember even the simplest problems.

Timed tests often trigger intense stress, not only because of the content but also because of the fear of running out of time. This pressure can cloud judgment and hinder performance, even for those who thoroughly understand the material. The speed at which they must work becomes the focus, overshadowing comprehension and retention.

This pressure manifests physically for some with sweaty palms, a rapid heartbeat, and the all-too-familiar blank mind. These symptoms contribute to a cycle of anxiety that becomes associated with future math tests, perpetuating the belief that math is inherently unmanageable.

Public problem-solving presents another arena where math anxiety flourishes. Imagine a classroom where a teacher asks students to solve problems at the board. For students with math anxiety, this can feel like being thrust into the spotlight of a stage they never intended to perform on. The fear of public embarrassment looms, with peers observing and silently judging. The anxiety

Help Teens Overcome Math Anxiety

Practice − Self-Defeating Attitudes = Success • Success • Success ...

involves not merely getting the answer wrong but also facing the perceived judgment from classmates. It's a fear of exposure, of having one's inadequacies revealed for all to see.

Sometimes, a teacher returns a test with a brief note saying, "You need to try harder," without providing constructive feedback. Such comments can reinforce a students' belief that they aren't good at math, leading to a self-fulfilling prophecy.

Peer influence is another crucial factor in classroom dynamics. Adolescents are highly aware of their peers' perceptions, and this awareness can significantly affect their confidence in math. Those with math anxiety may feel inadequate. They believe strongly that they will never measure up to their peers in a competitive classroom environment where student scores are frequently compared.

Peer comparison can be a double-edged sword; it may motivate some students to improve, but it can also reinforce feelings of inadequacy for others. When students perceive themselves as less capable than their peers, this perception can lead to a decline in self-esteem and further withdrawal from math-related activities.

Traditionally, many teachers emphasized rote memorization, requiring students to recall formulas and procedures without fully understanding the underlying concepts. This learning method can be stressful for students who prefer to **understand the underlying reasons behind the process**.

Without deeper comprehension, math becomes a series of disconnected tasks to memorize rather than an interconnected web of concepts to explore. This approach can alienate students who might otherwise excel if they could engage with the material in a more meaningful way.

Classroom Triggers

> ☐ Observe students' body language during timed tests. Do they show signs of stress?
> ☐ Monitor reactions when students are asked to solve problems publicly. Are they hesitant or visibly anxious?

This fear often leads to avoidance behaviors, such as students not raising their hands or volunteering for public problem-solving activities. Over time, this avoidance can create gaps in learning, as they miss valuable opportunities to engage meaningfully with the content. As they begin to choose a career, it may hinder their participation in courses essential for their chosen field.

The avoidance of math-intensive subjects extends beyond the classroom. It shapes students' perceptions of potential career paths, particularly those that require strong mathematical foundations, such as engineering, computer science, and economics.

Help Teens Overcome Math Anxiety

Practice – Self-Defeating Attitudes = Success • Success • Success ...

Students who shy away from math due to anxiety often show a decreasing interest in STEM fields. This is a loss not only of individual potential but also for industries that thrive on diverse perspectives and approaches. As math anxiety persists, it seeps into career choices, steering students away from professions where they might otherwise excel, were it not for the fear and discomfort that math induces.

This avoidance creates a feedback loop where anxiety contributes to performance issues, which in turn reinforces the anxiety. It's a cycle I've observed too frequently: a student dreads failing a math test and performs poorly because of anxiety, which intensifies their fears about future math challenges. This "fear-failure" cycle traps students in a pattern where each setback deepens their aversion to math. Breaking this cycle requires dismantling the fear component, a process that starts with **understanding** and addressing the root causes of the anxiety. Encouraging a **focus on learning and comprehension** rather than performance alone can be beneficial.

Teacher Journal

Consider how you can take immediate action to support students who show early signs of math anxiety. You may want to re-consider these at various points throughout the year.

- Have you noticed any physical or behavioral changes in students during math activities?
- How can you initiate a conversation with students about their feelings toward math?
- What strategies can you implement to create a more supportive environment for learning math?

Help Teens Overcome Math Anxiety

Practice − Self-Defeating Attitudes = Success • Success • Success ...

Chapter 2: The 5 Strategies

Starting off on the right foot!

Begin by creating a **nonjudgmental space** where students feel safe expressing their feelings. You might ask, "I've noticed that you seem stressed during math class. How are you feeling about the material?" This simple question can be a lifeline, offering students the opportunity to share their experiences.

Active listening is essential to validate students' feelings and reassure them that it's normal to feel anxious. It's crucial to convey that they are not alone in their struggles and that support is available. We discussed how we could all practice good listening skills in groups, during class, and outside of class.

Proactive communication serves as our most powerful tool that opens the door to understanding and intervention. However, starting a conversation about math anxiety can be challenging. It's important to be empathetic and open to all discussions. This helps students feel valued and important for them to share their ideas and concerns.

Consistency and collaboration can **build a supportive home and school environment**. In school, we can create an environment where students are encouraged to ask questions and view mistakes as a vital part of the learning process. Use **positive reinforcement** to acknowledge effort and improvement rather than just correctness. Parents can reinforce this

Help Teens Overcome Math Anxiety

Practice – Self-Defeating Attitudes = Success • Success • Success ...

support at home by engaging in positive discussions about math, perhaps by sharing stories of their own challenges and triumphs.

Some ways to build this environment could include active listening, proactive communication, empathy, and openness, as well as preventing escalation, utilizing relaxation techniques, and employing differentiated instruction with visual aids, hands-on activities, and technology.

Teachers can influence students' perceptions of their abilities through their **expectations and feedback**. High expectations can motivate students but may also intimidate them if they lack support and encouragement. A teacher's belief in a student's potential can be a powerful catalyst for success.

However, if a teacher shows doubt or frustration, it can deepen students' anxiety and diminish their confidence. Writing a brief note on papers to encourage students to review their work and improve their understanding of the content is something we can all do to help. Simply showing that something is wrong may discourage students and create more anxiety because they may be unaware of how to learn from the mistakes they make. Sometimes it can be valuable to do some error analysis as an assignment to encourage students to become better at reviewing their own work.

Student Journal #1

- What emotions do you feel when faced with a math problem?
- How does math anxiety affect your interactions with others in class?
- Have you noticed any broader stress that may impact your stress for doing math?

To prevent the escalation of math anxiety, we should implement both **proactive** and **responsive** strategies. One effective approach is to incorporate **relaxation techniques** into daily classroom routines. Simple practices, such as deep breathing exercises before a test or visualizations of a peaceful place, can help students manage their physiological responses to stress. Another strategy is to use **differentiated instruction** to meet students at their individual levels. This involves tailoring lessons to accommodate various learning styles and paces, ensuring that all students can access the material in a way that resonates with them. **Visual aids, hands-on activities, and technology** can all contribute to making math more accessible and less intimidating.

There are many facets of each strategy that can be supported. These strategies are also interconnected and mutually support one another.

The 5 Strategies

Help Teens Overcome Math Anxiety

Practice – Self-Defeating Attitudes = Success • Success • Success ...

> 1. **Mindshifts for Success**
> 2. **Mathematical Practices**
> 3. **Problem-Solving & Critical Thinking**
> 4. **Reflection and Reinforcement**
> 5. **Interventions**

2.1 Mind Shifts for Success

Growth Mindeset

Imagine a classroom filled with eager students who have unique dreams and aspirations, yet the invisible chains of math anxiety hold some back. These students aren't just struggling with numbers; they're battling a mindset that tells them they're not capable. It's a narrative I've encountered many times, and it's worth discussing.

Let's take a moment to consider the stories students tell themselves about math: "I'm just not a math person," "Math is only for geniuses," or "Math is my worst subject in school." These are more than just phrases; they're deeply ingrained beliefs that shape how students perceive their ability to succeed in mathematics.

Student Journal #2

- Think about a time when you told yourself you could not do something you wanted to do. How did it turn out?
- When someone did not believe you could do something, what did they say to you? How did their comments affect you and your performance?
- When you think about your mathematical journey, what regrets do you have?
- How would you like your relationship with mathematics to become?

The **concept of a growth mindset**, introduced by Carol Dweck, provides a way to shift these narratives. According to the concept, people can develop ability and intelligence through dedication and hard work. In mathematics, a growth mindset means seeing challenges as chances to learn and grow, not as problems to avoid. This mindset sharply contrasts with a fixed mindset, where students may see their math skills as static and unchangeable. Dweck's research highlights the transformative power of a growth mindset, demonstrating that when students believe they can improve, they often do. This belief fosters resilience, perseverance, and a willingness to embrace challenges—essential traits for success in math and beyond.[xi]

Help Teens Overcome Math Anxiety

Practice – Self-Defeating Attitudes = Success • Success • Success ...

Confidence

> Remember Tom's anxiety about numbers? We worked together to reframe his thoughts, helping him see math not as an insurmountable obstacle but as a puzzle to solve with patience and practice. Slowly, Tom's confidence grew, and so did his scores.

Building student confidence in math begins with promoting **positive self-talk** and encouraging practices that foster a **constructive internal dialogue**. Urging students to replace negative thoughts with affirmations like "I am capable of learning math" or "Mistakes help me grow" can significantly enhance their confidence.

Visualization exercises, where students imagine themselves successfully solving math problems, further reinforce this positive mindset. It's about cultivating a mental environment where students feel empowered rather than defeated.

Consider activities that enhance confidence, such as **collaborative group work** where students can share ideas and learn from each other. These activities improve understanding and foster community and support.

Emphasizing the values of **perseverance, persistence, and resilience** in mathematics is essential. Encourage students to appreciate returning to previous steps when encountering challenges or breaking down complex problems into smaller, more manageable parts. This strategy helps them recognize that challenges are part of the learning process and not a reflection of their abilities.

Classroom Environment

Creating a supportive classroom environment is essential for transforming math narratives. This includes debunking common myths about math ability and fostering a culture where every student feels valued and included. The classroom atmosphere should encourage engagement, open dialogue, and create trust.

Regular use of **positive reinforcement** helps students feel appreciated and motivated. Arrange seating to promote interaction and use **inclusive language** and materials that resonate with diverse learners. It's about building strong relationships between teachers and students, as well as among peers, where regular check-ins and feedback are the norm.

Recognizing and addressing math anxiety in teens is essential for supporting their academic and emotional well-being. Educators and parents play a vital role in identifying early signs, such as physical symptoms like sweaty palms or behavioral changes like withdrawal and responding with empathy. By shifting classroom dynamics from competition to **collaboration** and fostering open communication, we create a **supportive environment** that builds students' **confidence,**

Help Teens Overcome Math Anxiety

Practice − Self-Defeating Attitudes = Success • Success • Success ...

resilience, and curiosity; enabling them to conquer anxiety and succeed in math.

To help students manage math anxiety effectively, it is essential to encourage the development of coping mechanisms such as **mindfulness and relaxation techniques**, which can help calm the mind and body before tackling math problems.

Educators need to **foster a collaborative classroom environment** where students are encouraged to work together and **support** each other. It can help students feel safe expressing their concerns and seeking help from teachers, parents, or peers for addressing both the emotional and academic challenges associated with math anxiety.

Techniques like deep **breathing** and **mindfulness** enhance focus and reduce stress, allowing students to approach math calmly. By fostering resilience and confidence, we can empower students to succeed in math and all aspects of their lives.

Recognition

Recognizing individual and group achievements within the classroom community is a powerful motivator. Monthly math awards and showcases of student work foster pride and a sense of accomplishment. These practices allow students to see math as a collaborative journey, where they celebrate success and share the learning experience.

Strategies for breaking this cycle involve **reshaping the narrative** around math. Encouraging a perspective that values **effort** and **exploration** over perfection can open doors to new ways of thinking about mathematics. Emphasizing the process rather than the outcome can help students perceive math as less of a threat and more as an opportunity for learning. Introducing low-stakes assessments, where grades are less important, allows students to engage with math in a safe space without the fear of immediate consequences. Such approaches, when consistently applied, can gradually rebuild students' confidence.

By adopting these strategies, students can view math not as a source of anxiety, but as an opportunity for growth and discovery. It's about reshaping the narrative and crafting a mathematical journey rich with potential and possibility.

Help Teens Overcome Math Anxiety

Practice – Self-Defeating Attitudes = Success • Success • Success ...

Student Journal #3

- Before this class, when did you experience conversations about the math you were learning? Explain how it went and what.
- What were the benefits/deterrents you experienced in this class when discussing math?

Remind Students

- ☐ Recognizing that feeling anxious about math is normal. This is the first step toward overcoming it.
- ☐ It's crucial not to let anxiety prevent you from attempting problems.
- ☐ Replace negative thoughts about your mathematical abilities with encouraging affirmations.
- ☐ A positive attitude, free from negative self-talk, sets high expectations and can help you rise to the occasion.
- ☐ Remember to take advantage of the supportive community in your class; where confidence and preparation lead to success.
- ☐ Use deep breathing exercises to manage anxiety during math tasks.
- ☐ Remind yourself that "Each mistake is a step toward understanding," and "I can

> approach math with curiosity and confidence."

Help Teens Overcome Math Anxiety

Practice – Self-Defeating Attitudes = Success • Success • Success ...

2.2 Mathematical Practices

"Here we go again! Another math class." A comment from some students as they enter their math classroom with a defeated, negative attitude. It's our job to help them see the class as something positive, that they can do, and there is no reason to be fearful.

It's not just about numbers; it's about confidence, or sometimes, the lack of it. Many students find themselves trapped in a cycle of self-doubt regarding math, and this loss often stems from past experiences. Perhaps they encountered failure or felt overwhelmed by the rapid pace of learning. These experiences can instill a fear of failure that becomes a significant obstacle. Consistent practice and support can help develop confidence in math, just like any other skill.

Example: Homework

One evening my daughter was sitting at the kitchen table doing her homework. She was getting upset and not doing her work. She turned to me and said, "Why do I have to do all these problems? I already know how to do this and there are 20 problems on this page!"

"That's what your teacher has asked you to do. Let's see what you're doing," I said to her.

When I looked at her paper, I saw she hadn't even started. It was a full page of long division with 2-digit divisors. I walked her through the first problem by asking what she needed to do and then told her, "You're doing just fine! It looks like you understand what you're doing.

She cried, "But, I have to do all 20 of these! Why do I have to do so many?"

I felt so sorry for her, because I didn't think she needed to do all that work either. I let her use the calculator to check her work when it was done.

"But my teacher won't let us use calculators!"

"If you need to correct something, then you will be able to know it's right."

She immediately did them and checked her answers. I saw her fixing a couple of problems she missed and then she was done. Her use of the calculator was a vauable tool for her understanding.

Help Teens Overcome Math Anxiety

Practice – Self-Defeating Attitudes = Success • Success • Success ...

Daily Practice

Daily math practice is essential, whether through classwork or homework. Encourage students to look at it as building a house; each day's effort adds another brick to their foundation. Homework isn't merely an obligation; it's an opportunity to reinforce what they've learned in class.

When teens get confused, encourage them to **ask questions**. Refer to other math texts for different explanations. They could use flashcards for quick reviews or explore online resources. Joining a study group could also be beneficial because discussing problems with peers often provides new insights. If the school offers tutoring services, encourage students to take advantage of them. These resources are available to support all students to enhance their understanding and confidence.

Useful Mistakes

Mistakes are valuable in the story of learning mathematics. Presenting them as valuable learning tools rather than failures, shifts the narrative from self-doubt to **growth**. Promoting conversations and peer feedback sessions about problem-solving approaches and mistakes in the classroom encourages a **growth mindset**.

Students learn to see errors as steppingstones to understanding rather than pitfalls to avoid. Classroom activities and discussions about common math errors and their solutions offer valuable **learning opportunities**. They enable students to examine where they went wrong and how to correct their mistakes, transforming their self-doubt to a growth mindset. This approach fosters **resilience** and empowers students to face math challenges with **confidence**.

Encourage students to **break down complex problems** into smaller, more manageable parts. This method can transform a daunting task into something more achievable. Urge students faced with a page of math problems to concentrate on one step at a time instead of feeling overwhelmed. This strategy streamlines the task and builds confidence as they progress.

Student Journal #4

- What has been your most challenging math experience?
- How do you react when you make a mistake in math?
- How confident are you in doing math problems?
- What things could you think about that might help you become more confident?

Help Teens Overcome Math Anxiety

Practice – Self-Defeating Attitudes = Success • Success • Success ...

Journaling

Encourage students to keep a **math journal** to track their progress, document challenges, and celebrate achievements. Have them reflect on which strategies were effective and where they encountered difficulties. This can **reinforce learning** and provide a record of their **growth over time**. As they journal about their math journey, they will **identify patterns** and areas for potential improvement. It is a personal space to explore their relationship with math, free from judgment and pressure.

> My students wrote in their journals as a warm-up. We discussed them later in the lesson, at least once or twice a week. Students would share for a minute or two in their groups or with the whole class. This was my way of regularly reinforcing the ability to conquer math anxiety.

Resilience

Resilience refers to the ability to recover from setbacks and is essential for overcoming math anxiety. Start by introducing **incremental challenges**, gradually increasing the difficulty of problems to help students build persistence and adaptability. Complex math puzzles and activities are excellent resources as teachers encourage students to engage deeply with the material, testing their limits and expanding their comfort zones.

Remember, resilience isn't about never failing; it's about **learning from those failures** and continuing to move forward.

Learning Styles

Recognizing that students learn in different ways is essential for effective teaching. **Learning style theories** suggest that individuals have preferences for how they receive and process information.

Developing **multi-sensory math activities** accommodates various learning styles that enhance comprehension and retention. For kinesthetic learners, manipulatives like algebra tiles or interactive objects help make abstract concepts more tangible. Visual aids, such as diagrams and infographics clarify complex ideas and provide a visual representation of data, making it easier for visual learners to grasp. Creating effective visual aids involves thoughtful use of color and layout to highlight key concepts and relationships. Incorporating these elements into lessons can significantly boost understanding and engagement.

Recognition

A **classroom gallery** showcasing student-created visual aids is an excellent way to show understanding. It celebrates creativity and serves as a resource for peers to learn from each other's work.

Classroom displays showcasing student progress, such as a "Math Wall of Fame," not only celebrate individual achievements but also foster a sense of

Help Teens Overcome Math Anxiety
Practice – Self-Defeating Attitudes = Success • Success • Success ...

community and instant success. **Regular recognition** of individual and group achievements reinforces positive outcomes and encourages ongoing effort. By acknowledging progress, students feel valued and motivated to continue to learn the math.

Collaboration

Example: My Classroom

> Upon entering my class, students appeared to be seated in straight rows. However, I organized them into groups by having them turn their desks a quarter turn to form groups of four. This arrangement allowed students to collaborate. They would return their desks to rows at the end of class.

Some students hesitate to ask questions because of a fear of judgment or embarrassment. However, asking questions is a sign of strength, and often, others have the same question, but didn't ask. **Student-led math discussions and peer collaboration** provide numerous benefits.

I often encouraged my students to have discussions where they learn to articulate their thoughts and listen to their peers. This practice enhances understanding and engagement while developing critical thinking and communication skills.

Example: First Week of School

During the first week of school, I spent one day playing a cooperative game with squares (approximately 15-20 minutes). This activity had students seated in groups of five. Nothing else was allowed on their desks. I handed each group five different envelopes. They each had pieces of squares. The group's goal was to build five squares together as a team. They could not speak, but they could politely exchange or move pieces using hand signals or nods to achieve their goal. No envelope contained a complete square, so students needed to turn to others in the group to accomplish the task.

After the activity, we spent a few minutes discussing what it means to work as a team. I asked several questions, including: How did you feel when your square was completed, but the group could not finish the other squares? The team could only complete the task by you breaking down your square and sharing pieces with other team members. Each person needed to collaborate with others to complete the group activity.

This fostered a high level of interdependence among team members. Each member needed to be willing to contribute, and sometimes, this

Help Teens Overcome Math Anxiety

Practice – Self-Defeating Attitudes = Success • Success • Success ...

> meant being prepared to take risks and build mutual trust.
>
> We discussed how this serves as a model for how we should collaborate in our groups to accomplish our work in this class.
>
> When doing group work, I encouraged students to collaborate and then ask me when they all had the same question. This approach allowed me to circulate among a class of 36 students by visiting each of the nine groups. They also learned from and relied on each other, rather than always depending on the teacher. This strategy was invaluable when I needed to be absent.
>
> It created a clear advantage for substitutes, who consistently provided me with glowing reports about how diligently the students worked on their lessons. Because students were comfortable working in groups, they could continue their learning and support each other in their group. They could get their work done and not feel they had lost the day.

Collaborative learning with clear objectives for discussions that focus on activities boosts motivation and helps students appreciate the value of teamwork and cooperation. Many teachers assign roles, such as moderator, note-taker, and presenter, within groups to ensure all students participate and contribute.

Collaborative projects with shared goals and real-world applications can be effective. Encourage students to apply math concepts to practical situations to deepen their understanding and appreciation of the subject.

Incorporating audio resources for auditory learners and setting up interactive math stations with varied sensory inputs can cater to different learning preferences. Stations offer students the opportunity to engage with math in a hands-on, exploratory manner.

Example: Collaborative Groups

> I often introduced collaborative projects that allowed students to work together, share ideas, and learn from one another. These projects not only reinforced math concepts but also fostered a sense of community and belonging.
>
> Sometimes, I gave extra credit points to each member of a group when they achieved the highest average on a quiz or test. This encouraged their group collaboration so they could earn the extra points. (They didn't really amount to much but gave them the incentive.)
>
> By creating an environment where students felt supported and valued, I established a foundation of trust that empowered them to take risks and embrace challenges.

Help Teens Overcome Math Anxiety

Practice – Self-Defeating Attitudes = Success • Success • Success ...

Technology

Technology platforms like *Amplify Desmos Math* enable students to **visualize mathematical concepts** in real-time. Also, interactive whiteboards and features such as digital quizzes and mind maps create dynamic lessons that facilitate deep exploration and understanding of math.

By **integrating technology** thoughtfully, educators can create a rich and engaging learning environment that supports diverse learners and encourages curiosity and exploration. These collaborative environments encourage accountability through peer-assessment rubrics, which foster self-reflection and provide valuable feedback. They also increase engagement using interactive digital tools.

> I recently had the pleasure of observing student teachers using various platforms with their students. The students appreciated the integration of technology in their lessons and how it enhanced their mathematics learning.

Student Journal #5

- What technology have you used recently to support your math learning?
- How has technology influenced what you have learned in doing math?

Explore Digital Tools for Math

- ☐ There are many popular technology tools that enhance math learning.
- ☐ *Amplify Desmos Math* is one platform to support students with graphing.
- ☐ Interactive whiteboards facilitate dynamic math lessons.
- ☐ Visual learning can be enhanced through various programs.
- ☐ Interactive quizzes provide real-time student feedback.
- ☐ Students might create their own visual representations or digital mind maps.
- ☐ Auditory learners will appreciate the incorporation of audio resources.
- ☐ Some programs provide varied sensory inputs for use in interactive math stations.

2.3 Problem-Solving and Critical-Thinking

Navigating the maze of math problems can feel overwhelming for many students. It's not just about finding the correct answer; it's about understanding the journey to reach it. Students typically face challenges in math problem-solving due to misinterpreting the problem. Imagine reading a complex word problem and feeling lost in translation. The **language of math** can sometimes seem foreign, with phrases that obscure rather than clarify. Many students struggle to apply abstract concepts to real-world problems. It's as if math

Help Teens Overcome Math Anxiety
Practice – Self-Defeating Attitudes = Success • Success • Success ...

exists in a vacuum, isolated from reality, creating a disconnect that makes problem-solving appear insurmountable. The difficulty in bridging the gap between theory and practice is a common obstacle in mathematics education.

Word Problems

Students often wonder, "When will I ever use this in real life?" Without a clear connection to practical applications, math becomes a chore and makes the subject seem abstract and irrelevant instead of a useful tool. Integrating real-world problems into the curriculum helps students recognize the value of their learning and reduce anxiety by making math more relatable and applicable. [xii]

Example: Look for Real-World Problems

> I searched for realistic problems for my students to solve. By integrating them through the year, students became very productive at solving problems they may have encountered in the past or would likely face in the future.
>
> I also had a poster on my wall from Dale Seymour's publication of the poster "When Are We Ever Gonna Have to Use This?" Many times, I would see students going to the poster to see what math they would need for their future careers.

Budgeting and personal finance are excellent examples to show the practicality of math and equip students with valuable life skills. Geometry, for instance, is not merely a series of shapes and theorems; it serves as the foundation for architecture and design. It's also a foundation for using logic. Students understand the subject better by exploring how geometry is used in building and product design.

Teaching **problem-solving as a process** emphasizes the significance of understanding the journey. Shifting the focus from simply finding the correct answer to understanding the process enhances the learning experience. Structured methods like Polya's four-step approach to problem-solving provide a framework for students. [xiii] His method encourages students to understand the problem. This comprehensive approach not only helps in solving the immediate problem but also equips students with a flexible method applicable to future challenges.

Polya's Method

1. **Understand** the problem by breaking down the components and determining what's being asked.
2. **Devise** a plan by strategizing and selecting the best approach.
3. **Implement** the plan by executing the chosen strategy.
4. **Reflect** on the process and evaluate it.

Help Teens Overcome Math Anxiety

Practice – Self-Defeating Attitudes = Success • Success • Success ...

Activities aimed at strengthening resilience can be incorporated into everyday learning. Motivate students to ask questions and delve into concepts beyond the surface level through an inquiry-rich environment.

Open-ended math problems offer multiple approaches and solutions, nurturing creativity and critical thinking. Student-led discussions can promote ownership of learning and peer collaboration.

Example: How to Read Word Problems

> I would often tell my students, "In what job will you have a page of math problems to calculate?" The real world relies on you to interpret the information in a real situation and figure how to solve the problem. These are the word problems we practice in class. It is how and where the math is used."
>
> "Word problems are the most important part of learning math. We can't read them like we would read a novel. Each phrase will offer information about the situation. It is your job to figure out what they are giving you and what they want you to figure out."
>
> When reading word problems, I would mimic reading a problem as if it were part of a novel. Then I would step away and tell them, "I have no

idea what they want me to do with this information."

Then I read it again, but I would stop after each phrase and ask, "What does this mean." I continued this process, one phrase at a time and made notes on the board to solve a word problem.

"You may want to create a chart to organize the information or create an equation that will solve the problem. This will provide the information to solve for the answer to the problem."

Teaching students to follow this process can support them to manage most problems they encounter.

Example: Why a Table?

Valerie was a good student in my Algebra II class. She stopped by to get help after school because she was struggling with a model I used in class. Below is a sample problem of what we were doing in class.

Valerie was anxious about not knowing which numbers should go in which box. She dropped by after school one day, and we went through a problem; she was fine. She just had no idea why I drew the box on the board or how I filled in the information.

I realized she and many others probably didn't understand this was just one of the graphic

Help Teens Overcome Math Anxiety

Practice – Self-Defeating Attitudes = Success • Success • Success ...

> organizers I used. From that point on, I made it clear that there are many ways to approach problems, and I have found different ways to organize the information so I could solve the problem.

Example: Sample Problem

> An airplane travels to its destination, 2,910 miles with the wind in 3 hours. It then travels 7,140 miles against the wind in 6 hours. What is the rate of the airspeed and the wind for each trip?

	D	=	r	*	t
With wind	2910		$r + w$		3
Against wind	7140		$r - w$		6

> This is a distance problem that uses the equation: Distance equals Rate times Time ($D = R * T$), sometimes called the "dirt" equation.
> I wrote the equation on the board and drew a box to organize the information. I used (r) for the airspeed and (w) for the speed of the wind.

When students see how math concepts apply to everyday situations, the subject becomes less abstract and more engaging. Long-term projects that involve solving real community issues are also impactful. They provide students with the opportunity to apply math concepts in meaningful ways and foster a sense of purpose and connection to their community.

I encouraged my students to view problem-solving as an exploratory process rather than a race to the finish line. It's about cultivating curiosity and encouraging students to ask questions like, "What if I tried this method?" or "How would this concept apply in a different context?"

Fostering an environment that values exploration over speed, students learn to appreciate the nuances of math and develop critical thinking skills. This mindset not only enhances their problem-solving abilities but also cultivates a lifelong love of learning. Students can develop a deeper understanding of math and build the confidence to tackle even the most challenging problems in most areas of their life.

Example: Teaching Formulas

> I do not like memorizing and taught my students how to develop formulas from understanding the figure. Some students would learn songs or sayings that would help them remember a formula.

Help Teens Overcome Math Anxiety

Practice – Self-Defeating Attitudes = Success • Success • Success ...

> When teaching area and volume, I would teach my students the basics and then showed them how to develop the formula they needed.
>
> For example, when finding the area of 2-dimensional figures, I would start with the area of a basic rectangle, area equals length time width ($A = l \cdot w$) or area equals base time height ($A = b \cdot h$). This was the one I memorized and used it to determine many others, by looking at how they relate to a rectangle.
>
> When finding the area of a triangle, you can draw the related rectangle and take half of it. The square is the rectangle with base and height being the same measurement. A rhombus or parallelogram also uses $A = b \cdot h$ where you can cut off the slanted part and slide it over to the opposite side to form a rectangle.
>
> Formulas for 3-dimensional objects can be determined using similar strategies by starting with the rectangular solid. The basic formula for volume begins the formula *Volume* equals the *Area of the base* times the *height*. From there you can determine the volume of other figures, such as a cube, a pyramid, etc.
>
> These are just a few examples. There are many others that can be developed.

2.4 Mathematics Interventions
Student Journal #6

> - Have I been to an intervention to help me with mathematics?
> - How has it influenced my understanding of math, and what impact has it had on my learning?

Early intervention is crucial in preventing long-term academic consequences. Recognizing the signs of math anxiety early allows educators and parents to provide support before patterns of avoidance become entrenched. For instance, a school-wide initiative could involve training teachers to identify anxiety symptoms and employ classroom strategies to mitigate them.

Many students learn from their classmates who may provide a role model of someone who may have shared their struggles in the past but since overcame them.

Help Teens Overcome Math Anxiety

Practice – Self-Defeating Attitudes = Success • Success • Success ...

Early Intervention Strategies

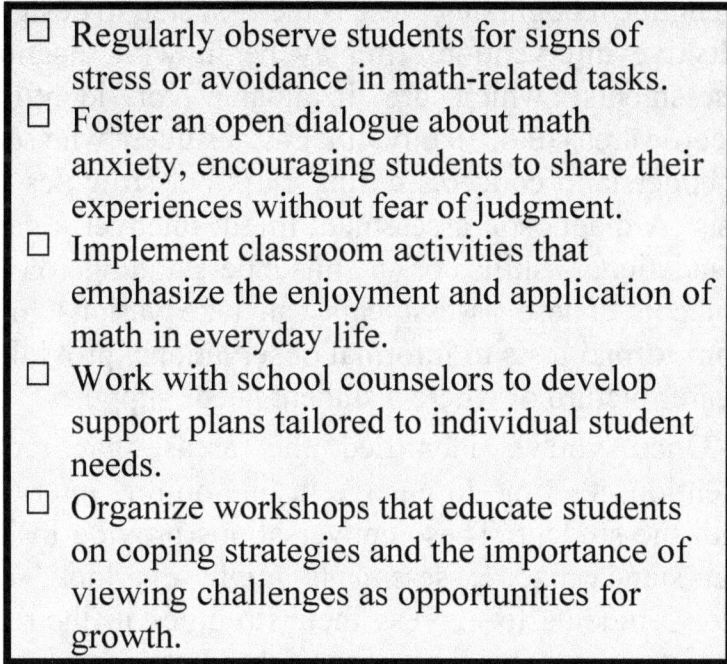

- ☐ Regularly observe students for signs of stress or avoidance in math-related tasks.
- ☐ Foster an open dialogue about math anxiety, encouraging students to share their experiences without fear of judgment.
- ☐ Implement classroom activities that emphasize the enjoyment and application of math in everyday life.
- ☐ Work with school counselors to develop support plans tailored to individual student needs.
- ☐ Organize workshops that educate students on coping strategies and the importance of viewing challenges as opportunities for growth.

Addressing math anxiety involves more than just improving scores; it opens doors to future possibilities. It focuses on turning the tide early before the anxiety takes root too deeply. By understanding the significant impact of math anxiety on academic performance, educators and parents can take proactive steps to support their students. Together, through **understanding and intervention**, we can assist students toward a future where math becomes a tool for learning rather than a source of fear. **Personalized and positive feedback** can reassure students, helping them feel valued and capable, even when faced with challenges.

Each student presents a unique set of challenges and strengths; recognizing these is the first step in designing effective interventions. It may begin with diagnostic assessments, which are invaluable for identifying specific areas of difficulty. Imagine a student who seems to understand concepts during class but struggles with tests. A diagnostic assessment might uncover a gap in foundational skills or a misunderstanding of key concepts. These assessments can take various forms, from formal tests to informal observations, providing a clearer picture of where a student needs support.

Once you've identified the areas that require attention, it's time to engage in one-on-one interviews with the student. These conversations provide insights that standardized assessments might overlook. They allow students to express their struggles in their own words, giving a glimpse into their experiences and feelings about math. You might discover that a student is grappling with anxiety from experiences or that they don't perceive the relevance of math in their lives. Understanding these personal narratives is essential for shaping an intervention plan that resonates with the student. It's about building trust and demonstrating to students that their voices matter in their educational journey.

With a **comprehensive understanding** of student needs, you can start crafting structured intervention

Help Teens Overcome Math Anxiety
Practice – Self-Defeating Attitudes = Success • Success • Success ...

plans. These plans should encompass short-term and long-term objectives tailored to each student's unique profile.

Short-term goals focus on mastering specific skills or concepts, while long-term goals aim to improve math confidence and performance. Incorporating student interests into these plans can significantly enhance engagement. For instance, if a student loves sports, like soccer, softball, football, or basketball; you might use a diagram of the court or field to compare various sports activities or statistics from their favorite team to teach **mathematical concepts**. Connecting math to their passions makes learning more relevant and enjoyable.

Set SMART Goals

Specific goals clarify what needs to be achieved.

Measurable goals allow for tracking progress.

Achievable goals keep students motivated and prevent them from feeling overwhelmed.

Relevance ties the goals to students' interests and real-world applications, making them more meaningful.

Time-bound goals create a sense of urgency and structure, encouraging consistent effort and progress.

This approach can prevent minor misunderstandings from becoming larger obstacles down the road. For students, the journey through math interventions involves active participation and commitment.

Encourage students not to hesitate to contact teachers, tutors, or peers to clarify concepts they struggle with. They should view these resources as allies in their learning journey. Having a study buddy or joining a group can provide additional support and new perspectives. If students miss a class, emphasize the importance of obtaining class notes from a fellow student. Missing important lessons can make it harder to catch up. Staying informed and engaged is crucial for maintaining momentum in their learning.

Remember each student is on their own path, with unique challenges and triumphs. The goal is not just to improve math performance but to empower students to **take ownership** of their learning and develop skills that will serve them well beyond the classroom. By providing **personalized support** and fostering an environment of encouragement and growth, you can help students unlock their potential and approach math with confidence and curiosity.

Share and discuss the following about interventions with students periodically.

Help Teens Overcome Math Anxiety

Practice – Self-Defeating Attitudes = Success • Success • Success ...

Math Intervention

- ☐ Involves regular practice to build proficiency and confidence.
- ☐ It's not about cramming before a test but about engaging with math consistently.
- ☐ Encourage students to dedicate time daily to focus on difficult areas, gradually increasing their confidence through repetition.
- ☐ Remind students that math is cumulative and builds on itself, so having a firm grasp of earlier material is important.
- ☐ If students encounter gaps in their understanding, reassure them that it's never too late to relearn foundational concepts.

2.5   Reflection and Reinforcement

Reflection plays a pivotal role in enhancing comprehension and retention of mathematical concepts. When students take time to reflect on what they've learned, they reinforce their understanding and identify areas where they need further exploration.

This reflection acts as a bridge between new knowledge and existing understanding, allowing students to consolidate their learning. It encourages them to **think critically** about their problem-solving approaches and recognize patterns in their thought processes that should be encouraged. By integrating reflection into daily math practices, students develop a habit of introspection that enhances their ability to retain and apply mathematical concepts effectively.

Students will begin to identify areas where they may require additional support. Students can **celebrate their growth**, which reinforces their confidence and motivation.

Various activities allow for integrating reflective thinking into math lessons. Encourage students to engage in end-of-lesson reflections, including journaling for their personal record of growth. These relections could consider what they've learned and the challenges they faced. Students can share their writing in small groups, which will foster a collaborative learning environment for students to learn from each other's experiences.

By creating a culture of reflection, educators can empower students to take ownership of their learning and develop the skills needed to succeed in math and beyond.

Student Journal #7

- What reflection strategies about learning math have you used?
- What has been your outcome, and how has it influenced your mathematics learning?

Help Teens Overcome Math Anxiety

Practice – Self-Defeating Attitudes = Success • Success • Success ...

Encourage students to set incremental goals based on their reflections of progress. By regularly engaging in this practice, students develop a deeper understanding of their learning process and gain insights into how they can improve and grow.

Embrace the opportunity to explore new strategies and approaches, knowing that each step forward brings you closer to mastering the subject. Through reflection and reinforcement, students can gain the confidence and skills needed to succeed in math and beyond, unlocking their full potential and paving the way for a bright future.

Teacher Journal

- What reflection strategies have you used to reflect on your teaching of mathematics?
- How has this influenced the outcome for your students?
- What strategies will you continue to incorporate in your classes?

School-wide initiatives can play a pivotal role in addressing math anxiety on a broader scale. Workshops and seminars that focus on math anxiety can be invaluable, offering both students and teachers new tools and perspectives. Encouraging a school culture that celebrates diverse talents and abilities, rather than just academic excellence, can shift the focus from competition to collaboration.

Schools might implement math clubs or after-school programs designed to explore math in a fun, pressure-free environment. These initiatives can foster a sense of community and shared purpose, reducing the stigma associated with math struggles and promoting a more positive approach to learning.

As we conclude this chapter, consider how these strategies align with the broader context of supporting students in math. The journey doesn't end here—it's just the beginning of a lifelong relationship with learning and discovery.

Reminders for Teachers

- ☐ Encourage peer collaboration rather than competition. How are your students interacting during group work?
- ☐ Evaluate your own feedback methods. Are they constructive and supportive?
- ☐ Integrate real-world scenarios into math lessons. Are students more engaged with practical applications?

Help Teens Overcome Math Anxiety

Practice – Self-Defeating Attitudes = Success • Success • Success ...

Chapter 3: Educators & Parents

3.1 Change Agents

You've likely seen it before—a student who struggles with math, not because of their ability, but because of a lack of confidence, support, or encouragement. Educators and parents can uniquely transform how students experience and engage with mathematics, serving as catalysts for change. As educators hold the torch of knowledge and methodology, and parents provide the supportive foundation, our roles are intertwined, enhancing a student's educational journey. Together, we can create an environment where math anxiety is not a barrier, but a step toward greater understanding and confidence.

The influence of educators in the classroom extends far beyond teaching formulas and solving equations. Teachers have the power to **ignite a passion for learning** and **foster a growth mindset** in their students. **Enthusiasm** for math can be contagious, inspiring students to view challenges as **opportunities to grow**.

By **creating an inclusive and supportive classroom** environment, teachers encourage students to **take risks** and **embrace their mistakes** as part of the learning process. It's about **setting high expectations** while providing the scaffolding students need to reach those heights. This **balance of challenge and support** is crucial for cultivating an atmosphere where students feel empowered to explore math without fear.

Parents play a crucial role in supporting their students' math education. Their involvement can significantly affect attitudes towards math, serving as both a motivator and a source of reassurance. It's important to remember that parents don't have to be math experts to make a difference. Provide encouragement and show interest in math activities that can go a long way. Simple actions, such as **discussing math homework** or exploring math-related topics together, reinforce the connection between school learning and real-world applications.

Building trust is fundamental for any successful partnership. Teachers should create an environment where parents feel comfortable sharing their thoughts and concerns. **Consistent positive interactions** build this trust over time. When parents see that teachers genuinely care about their child's well-being and success, they are more likely to engage and collaborate. Similarly, parents should **feel empowered to advocate** for their child's needs and to ask questions about the educational process. This mutual trust and respect form the foundation for a strong, productive partnership.

As an educator, remember that not all parents have the same access to technology or the same availability for meetings. Offer various communication options, such as email, phone calls, virtual meetings, or in-person conferences, can accommodate diverse needs and

Help Teens Overcome Math Anxiety
Practice – Self-Defeating Attitudes = Success • Success • Success ...

preferences. Being adaptable shows a commitment to collaboration and inclusivity, making it easier for parents to participate actively in their child's education.

Not all forms of involvement are equally beneficial. For example, studies have found that direct homework assistance can have a negative correlation with math performance. This may be because parents inadvertently take over the task, rather than guide their child through the problem-solving process. To avoid this, parents should provide support and encouragement. Guide students to resources and strategies that promote independent learning.

The cultural context also plays a role in parental involvement. In some cultures, there is a strong emphasis on academic achievement, which can significantly affect a student's motivation and performance. Understanding these cultural nuances can help educators tailor their approaches to meet the needs of diverse student populations. Acknowledge and respect cultural differences. In this way, teachers can build stronger relationships with parents and create a more inclusive learning environment.

Together, educators and parents can create an environment where students **feel valued and encouraged** to explore their potential. This partnership is a powerful tool for overcoming math anxiety and nurturing **a positive learning experience**. By working collaboratively, you pave the way for a brighter future for your students, where math is not just a subject to be endured but a field to be explored and enjoyed.

3.2 Encourage At-Home Math Activities

Picture a world where math extends beyond classrooms, enriching interactions and daily routines by becoming a part of everyday life. This vision is not only possible but also essential for reinforcing math learning at home. Parents hold a unique power to transform everyday activities into engaging math lessons. You don't need to be a math expert to make a **significant impact**. Sometimes, all it takes is a little **creativity and enthusiasm** to weave math into the fabric of daily life, making it a natural and enjoyable part of your child's experience.

Consider starting with cooking. It's an activity most families engage in, and it's teeming with math opportunities. While preparing a meal, invite them to double or halve a recipe. Cooking provides a chance to work with ratios and proportions when scaling recipes up or down. Ask them to measure ingredients using different units, such as converting between cups and tablespoons, to reinforce their understanding of measurement. These exercises not only teach math but also **build confidence** in applying math in real-world situations.

Beyond cooking, budgeting offers another rich avenue for math practice. Involve your child in planning

Help Teens Overcome Math Anxiety

Practice – Self-Defeating Attitudes = Success • Success • Success ...

a family outing or grocery shopping trip. Set a budget and let them help manage expenses, calculate costs, and compare prices. This activity enhances skills in addition, subtraction, and multiplication, as well as percentage calculations when considering discounts or sales tax. Discussing needs versus wants during this process can also introduce concepts of financial literacy and decision-making. Such practical experiences allow students to see the relevance of math in everyday decisions, fostering an appreciation for its value beyond the classroom.

Tasks like organizing a bookshelf or pantry by size or category introduce the concepts of sorting and classification. Measuring a room for new furniture or rearranging items can involve geometry and spatial reasoning. Even something as simple as timing how long it takes to complete different chores can prompt discussions about units of time and estimation. These activities help students practice math in a hands-on manner, which can be beneficial for those who learn best through tactile experiences.

Game-based learning is another powerful tool for reinforcing math at home. Board games, card games, and educational apps can make math practice fun and engaging. Games like Monopoly or The Game of Life introduce concepts of money management and strategic planning. Card games that involve counting, probability, or pattern recognition can sharpen mental math skills.

It's also important to encourage students to **set realistic goals** for their math learning. These goals should be **specific, measurable, and achievable**, allowing for clear progress tracking. For example, a goal might be to master a particular type of math problem by the end of the week. Celebrate their achievements, no matter how small, to motivate them to continue pushing forward. This practice not only builds math skills but also enhances self-discipline and perseverance, qualities that are valuable across all areas of life.

For students who need additional support, consider engaging a **math tutor**. A tutor can provide personalized assistance, helping your child find those "aha" moments of understanding. Tutors can offer new perspectives on challenging concepts and serve as a valuable resource for students who benefit from one-on-one attention and guidance.

Technology can also be a great ally in supporting math learning at home. Virtual courses and online games offer diverse learning methods that cater to different learning styles. Encourage students to explore educational websites and apps that provide interactive tutorials, practice problems, and quizzes. These resources can supplement schoolwork and offer opportunities for independent learning. Embracing technology in this way can make math more accessible

Help Teens Overcome Math Anxiety
Practice − Self-Defeating Attitudes = Success • Success • Success ...

and engaging, especially for students who enjoy digital platforms.

Remember, the goal of at-home math activities is not only to reinforce academic skills but also to foster a positive attitude towards math. By integrating math into daily life, parents help students see it as a natural and enjoyable part of their world. This approach builds confidence and curiosity, encouraging exploration of math beyond the confines of textbooks and classrooms.

Together, educators and parents can create an environment where math is not just a subject but a gateway to understanding the world. We should continue exploring these opportunities to pave the way for a future where everyone embraces math with enthusiasm and confidence.

It's Not About How You Teach, It's About How Your Students Learn

Help Teens Overcome Math Anxiety

Practice – Self-Defeating Attitudes = Success • Success • Success ...

Conclusion

Throughout this book, we've embarked on a transformative journey to reshape the way we approach math education. Our overarching goal has been to address the pervasive issue of math anxiety and cultivate a positive math culture among educators, teens, and parents. By delving into the strategies and insights shared in these pages, you've taken a significant step towards empowering yourself and those around you to embrace math with confidence and enthusiasm.

Let's take a moment to reflect on the five strategies we explored:

1. Mindshifts for Success
2. Mathematical Practices
3. Problem-Solving and Critical Thinking
4. Reflection and Reinforcement
5. Mathematics Interventions

Each of these strategies plays a vital role in fostering a supportive and engaging math learning environment. Each strategy is filled with many aspects to support all teens. By incorporating these approaches into your teaching practices and parenting techniques, you can make a profound impact on the way adolescents experience and engage with mathematics.

As we've discussed, the key to overcoming math anxiety lies in shifting the mindset. It's about embracing challenges as opportunities for growth and recognizing that mistakes are valuable steppingstones in the learning process. By fostering a growth mindset and encouraging a love for learning, we can help students develop the resilience and confidence needed to excel in math and beyond.

The Power to Make a Difference

> Throughout my 50 years of teaching, I've witnessed firsthand the transformative power of these strategies. I've seen students who once dreaded math class become eager learners, excited to tackle new challenges. I've watched as parents and educators collaborated to create supportive environments that nurture mathematical curiosity and exploration. These experiences have reinforced my belief that change is possible, and that each of us has the power to make a difference.

So, my **CALL TO ACTION** for you is to **actively apply the strategies and insights** you've gained from this book. Whether you're an educator, parent, or student, you play a crucial role in fostering a positive math culture. Start by reflecting on your own experiences and identifying areas where you can implement these strategies. Engage in **open and honest conversations** with your students, children, and

Help Teens Overcome Math Anxiety

Practice − Self-Defeating Attitudes = Success • Success • Success ...

colleagues about math anxiety and the importance of a growth mindset.

Create opportunities for hands-on learning, collaborative problem-solving, and reflection. Encourage students to take risks, embrace challenges, and learn from their mistakes. Celebrate their successes, no matter how small, and help them see the value in their efforts. By consistently applying these strategies, you can create a ripple effect that extends far beyond your immediate circle.

But the journey doesn't end here. As you continue to explore and implement these strategies, I encourage you to stay curious and open to new ideas. The field of mathematics education is constantly evolving, and there is always more to learn. Seek professional development opportunities, engage with fellow educators and parents, and stay committed to lifelong learning and improvement.

Together, we have the power to transform the way teens experience and engage with mathematics. By fostering a positive math culture and empowering students to embrace their potential, we can help them develop the skills and confidence needed to succeed not only in math but in all aspects of their lives.

Thank you for joining me on this transformative journey. I am confident that the strategies and insights you've gained will serve you well as you continue to make a positive impact in the lives of adolescents. Remember, change starts with each of us. We can create a world where math empowers and brings joy if we embrace this responsibility with open hearts and minds.

Help Teens Overcome Math Anxiety

Practice − Self-Defeating Attitudes = Success • Success • Success ...

Acknowledgements

I would like to express my heartfelt thanks to everyone who offered support and advice throughout the writing of this book. There are far too many to name individually, but each contribution has meant a great deal to me.

My sincere gratitude goes to **Pam Sheppard** who provided some great insights when I began writing this book. I also want to extend my gratitude to **Ellen Ballard**, who carefully edited the final manuscript. **Laura Brislawn** and **Pat Slusser** were especially valuable to my writing with their thoughtful feedback and attention to the mathematical details. Many others provided insights and suggestions that have found their way into these pages.

Finally, a very special thank-you to my past students, whose experiences, questions, and honesty provided the most meaningful inspiration for addressing math anxiety and how to overcome it.

It's Not About How You Teach, It's About How Your Students Learn

Help Teens Overcome Math Anxiety

Practice − Self-Defeating Attitudes = Success • Success • Success ...

About the Author

Dianne DeMille, PhD, is a retired educator with over 50 years in public school education. Her experience included teaching every math course in high school, including AP Calculus and grades 6-8 for 3 years during the middle of her career. She retired in 2012 as Mathematics Coordinator at the *Orange County Department of Education*, working with all schools in Orange County, California. During those years, she wrote several educational books for the County, *California State Department of Education*, and *California State University*. Dianne currently supports new high school math teachers in Orange County schools at *California State University, Fullerton*.

Since retiring, Dianne co-authored a memoir about her father, *It Started with A Pencil: Memoir, Leslie B. DeMille*. She also co-authored four True Crime books, *In the Furtherance of Justice* (formerly *Path of the Devil*) with Larry Hardin, Jeff Pearce, and Randy Torgerson, and the other three were written with retired DEA Agent, Larry Ray Hardin as writer of *Fighting My Greatest Enemy: Myself*, *Home is Never the Same*, and *Life's a Journey Between Heaven & Hell*. *The Life of Riley Living with Duchenne Muscular Dystrophy* was another memoir written with Riley's mother, Nina Stuart

Herrera. Most recently, Dianne supported Timothy Snodgrass and Larry Ray Hardin as editor and publisher of their fiction novel based on real facts, *Between the Devil and the Badge*. Visit her website, *diannesconsultantservices.com*.

References

Bohrod, Nina, Blazek, Candace, & Verkhovtseva, Sasha. (2011). *How to Overcome Math Anxiety* Retrieved June 12, 2025, from: https://www.weber.edu/wsuimages/vetsupwardbound/studyskills/overcomemathanxiety.pdf

David Gay stated in *Solving Problems Using Elementary Mathematics* Pearson College Div in 2016.

Dweck, Carol. (2008). *Mindsets and Math/Science Achievement.* Retrieved June 12, 2025, from: http://www.growthmindsetmaths.com/uploads/2/3/7/7/23776169/mindset_and_math_science_achievement_-_nov_2013.pdf

Education Northwest. (September 2019). *Mathematics Interventions: What Strategies Work for Struggling Students.* Retrieved August 9, 2025, from: https://educationnorthwest.org/resources/mathematics-interventions-what-strategies-work-struggling-students

Kogelman, Stanley & Warren, Joseph. (1978). *Mind over Math*, Dial, New York, 1978 McGraw-Hill Book Company, pp. 30-43

Luo, Yilei & Chen, Xinqi. (August 1, 2024). *The Impact of Math-Gender Stereotypes on Students' Academic Performance: Evidence from China.*

Retrieved August 9, 2025, from: htttps://pmc.ncbi.nlm.nih.gov/articles/PMC11355439/

Oviedo. (January 23, 2023*). Math in Action, Virtual Learning: 8 Ways to Improve Math Skills at Home - Mathnasium.* Retrieved July 15, 2025, from: https://www.mathnasium.com/math-centers/oviedo/news/8-ways-improve-your-math-skills-home-oviedo

Parker, Clifton B. (December 17, 2015). *Cultivating a growth mindset in mathematics.* Retrieved June 12, 2025, from: https://ed.stanford.edu/news/cultivating-growth-mindset-math.

Peabody Researchers (Jan 21, 2017) Article *Math Myths: Researchers Debunk Common Misconceptions*, researchers at Peabody. Retrieved August 13, 2025, from: https://news.vanderbilt.edu/2017/01/31/math-myths-researchers-debunk-common-misconceptions/

Saunders, Hal. (January 1, 1988). *When Are We Ever Gonna Have to Use This?* Dale Seymour Publications.

Schwartz, Sarah (2022). Pisa 2012 compared to 2022— November 25, 2024. Which Nation's Students Are Defying the Math Anxiety Trend? Retrieved September 2, 2025, from: https://www.edweek.org/teaching-learning/which-nations-students-are-defying-the-math-anxiety-trend/2024/11.

Szűcs, Dénes. (September 1, 2015). *Neural correlates of math anxiety – an overview and implications.* Retrieved July 15, 2025, from: https://pmc.ncbi.nlm.nih.gov/articles/PMC4554936/

Szűcs, Dénes. (September 15, 2019). *The origins of math anxiety and interventions.* Retrieved July 15, 2025, from: https://solportal.ibe-unesco.org/articles/the-origins-of-math-anxiety-and-interventions/

Tobias, Sheila (1978). *Overcoming Math Anxiety.* Boston: Houghton Mifflin.

Todd, Jeff. (January 9, 2024). *Building Productive Parent-Teacher Relationships for Math Learning.* Retrieved June 12, 2025, from: https://www.sadlier.com/school/sadlier-math-blog/building-productive-parent-teacher-relationships-for-math-learning

Vanbinst, K., Bellon, E., & Dowker, A. (2020). *Mathematics Anxiety: An Intergenerational Approach. Frontiers in psychology*, 11, 1648. Retrieved February 22, 2025, from: https://pmc.ncbi.nlm.nih.gov/articles/PMC7385133/

Wang, Xueshen & Wei, Yun. (Decembr 26, 2024). *The influence of parental involvement on students' math performance: a meta-analysis.* Sec.

Educational Psychology Volume 15. August 13, 2025, from: htttps://www.frontiersin.org/journals/psychology/articles/10.3389/fpsyg.2024.1463359/full

---

[i] Tobias,
[ii] Steinam
[iii] Tobias
[iv] Schwartz
[v] Ibid
[vi] Kogelman and Warren
[vii] Gay
[viii] Researchers at Peabody
[ix] Ibid
[x] Vanbinst
[xi] Dweck
[xii] Saunders

[xiii] Polya

www.ingramcontent.com/pod-product-compliance
Lightning Source LLC
LaVergne TN
LVHW020059090426
835510LV00040B/2641